CONTENTS

Continuing Professional Development

Visual amenity valuation of trees and woodlands

PREFACE

The methods described in this publication are based on proposals which were first published in the Arboricultural Journal (Helliwell 1967). The method of evaluating individual trees was subsequently adopted, in slightly modified form, by the Tree Council and published as a leaflet in 1974. This leaflet was re-printed and updated by the Arboricultural Association in 1984. A slightly modified version of the evaluation method for woodlands was also published as a leaflet by the Association in 1986.

Since its adoption by the Tree Council, the method for individual trees has been extensively used in court cases, insurance claims, and public inquiries and is to our knowledge the only method of valuation of the visual amenity of trees which has currently been accepted in British courts. It has also been used to place a value on trees within a neighbourhood (Dolwin and Goss, 1993). The method for woodlands has been used to a somewhat lesser extent, and has been used in court by the Forestry Commission.

The basic approach of the Helliwell System is to allocate scores under a number of different factors. These scores are then combined to give an overall comparative score for a tree or woodland. As a further step, it is then possible to attach a value to this score by use of a monetary conversion factor which has been developed since 1974 by periodic meetings of a working group under the auspices of the Tree Council, and which will be assessed again from time to time.

Other methods of valuation are available (see References). Most of these have a different approach, being based on the costs associated with purchasing, planting, and maintaining trees, with some adjustment for the location and condition of the tree. The Tree Council acknowledges the existence of such methods but they are not the subject of this publication.

The Tree Council and the Arboricultural Association are pleased to publish this revised edition of the Helliwell System, in which some further changes have been made in the light of experience and comments received. The Arboricultural Association has in recent years organised workshops on the use of this System under the guidance of Rodney Helliwell and Steve Coombes. This has provided additional feedback, which has been helpful in guiding the adjustments that have been made from time to time.

The recommended monetary conversion factors for trees and woodlands are updated in line with the Retail Price Index at the start of each year and are published by the Arboricultural Association on its website at www.trees.org.uk/publications.

Tree Council
2008

Continuing
Professional
Development

1. Introduction

This system was originally devised as a means of achieving logical decisions in the planning and management of woodlands and urban tree populations, by assessing the relative contribution of different trees and woodlands to the visual quality of the landscape. It was not devised as a means of attributing values to trees for compensation or for the calculation of appropriate fines, but it has been used for these additional purposes in numerous instances. It is inevitable that some of the factors involved are rather subjective, as people's perceptions of visual amenity are subjective. This can result in different users arriving at different values; but it is usually possible for such differences to be resolved, either by the users themselves or by an independent arbitrator.

The basis of this system is firstly to place trees with different characteristics into a rank order, so that the least valuable trees are at the bottom and those of greatest value at the top. There is then a need to ensure that the intervals on the scale are more or less even, so that a difference of one point on the scale is similar to a difference of one point elsewhere and the system is therefore internally consistent. If necessary, a monetary value can then be attributed to the points on the scale, by applying a monetary conversion factor, as discussed in Section 3.

It should be noted that the valuation of trees for amenity purposes can never be a totally precise exercise, and this booklet should be regarded as a **guide** rather than a precision instrument.

The methodology is fairly straightforward and simple to apply. However, the assessment of the condition and likely longevity of a tree will require a relatively high level of knowledge and experience in these matters. The other factors require less training but, where possible, valuation should be carried out by a person with a sound knowledge of arboriculture and who has had previous training and experience in the use of this method.

In most instances, valuation will be carried out on behalf of the general public or some other group of people who may not be able to tell an oak tree from an ash; although they probably appreciate leafy surroundings, even if only subliminally. In this context, the species or type of tree is only of importance insofar as it affects its size, longevity, and suitability for the location. Most trees register visually as green blobs in the landscape, and some people may not even consciously recognise the fact that trees are present; although even they are likely to recognise that areas with trees have a different "feel" about them. A recent study on responses to scenes with trees of different form (Lohr and Pearson-Mims, 2006) indicated that trees with a more natural spreading form evoked a greater positive emotional response than more formal conical or rounded forms, although all forms of tree evoked a stronger response than inanimate objects. It should also be noted that the effect of trees on the character of a locality is not always assisted by the use of trees with vivid flowers, odd shapes, or foliage of striking colour (Helliwell 1983) – the best dressed person is not always the one with pink socks, yellow shirt, and purple trousers! Similarly, a restrained use of trees with plain green leaves and a modest amount of blossom or autumn colour will often be more appropriate than the planting of numerous different varieties; particularly where trees are just one element in a landscape which also contains buildings, roads, and other features.

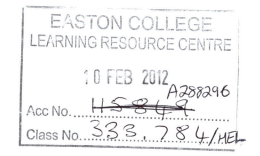

Continuing Professional Development

The objectives of evaluation may include:

(a) Provision of a structured approach to the allocation of resources for the planting and management of trees and woodlands.

(b) Provision of a system for assessing the loss of visual amenity when trees are accidentally or maliciously damaged or destroyed.

(c) Enabling advisors and planners to include trees and woodlands in their calculations with a similar degree of consistency to that which is possible for other aspects of design and management.

Visual amenity value, it must be stressed, may be counter-balanced by other factors. For example a farmer may have a compelling need to remove a tree in order to provide access to a field, which may be sufficient to override the amenity value of the tree. It should also be clearly stated that this system attempts to derive a visual amenity value only, and is not the whole case for or against any tree or woodland which is being assessed.

Other **benefits** provided by trees, such as timber, wildlife, shade, shelter, historical/cultural associations, climate control, and storm water control may also be important, and may need to be assessed, separately.

In addition, the **costs** involved in planting and maintaining trees and in precluding or affecting other uses of the land may need to be considered before any decisions are taken.

Table 1 lists some of the benefits that may be provided by trees, together with factors of relevance to those benefits:

Benefit	Factors	Bases for monetary valuation
Timber and other products such as fruit, foliage, etc.	Species, size, quality, quantity, accessibility, current demand	Market prices, local employment, maintenance of a sustainable resource, balance of payments
Visual amenity	Suitability, size, location, longevity, form, numbers of other trees	Effects on property prices, effects on health and levels of vandalism, tourism, and economic regeneration. **Independent** decisions on expenditure to retain or plant trees
Shade to buildings and open spaces, and glare reduction	Climate, type of building, type of tree, size, location, aspect	Reduction in cost of air conditioning, effect on property prices, cost of alternative engineering solutions
Shelter to buildings	Climate, height, permeability, number and distribution of trees, aspect	Reduction in heating costs, effect on property values, cost of alternative solutions
Pollution reduction	Type of pollutant, type of tree, location of tree and number of other trees	Property values, improved health
Flood reduction	Climate, landform, land management, type of tree or woodland	Reduction in flood control costs, or reduced damage to property, cost of alternative solutions
Carbon sequestration	Rate of growth, density of timber, uses of timber, amount of fuel used to transport and process the timber	Costs of alternative methods of reducing the amount of atmospheric carbon dioxide
Nature conservation	Type of tree, history and management of site, age and condition of tree, associated species	Contribution to conservation of nature
Heritage value	Tree history and associations	Contribution to leisure/tourism economy

It should also be clear that, in cases where a figure for compensation is being calculated, **compensation** may be paid **either** for the loss of amenity **or** the cost of its replacement, and **not for both** of these; and it should be stressed that the visual amenity value of a tree may be more or less than the cost of replacement. Only occasionally will it happen to be similar.

In many instances the cost of replacement would be considerably greater than the loss of amenity. Where large or old trees are involved, direct replacement will not usually be feasible, and it is the author's view that compensation for the loss of such trees can only be sensibly based on the loss of benefit which has ensued. The recipient may choose to use all or part of any compensation to pay for a new tree. It might be decided to plant a small tree at relatively little cost, in which case it will be some time before the amenity is replaced, but the surplus money could be spent on other trees or on other more pressing matters; or it may be decided to plant a somewhat larger tree at greater cost. But in any normal case it would not be logical to pay more in compensation than the loss of benefit which has occurred.

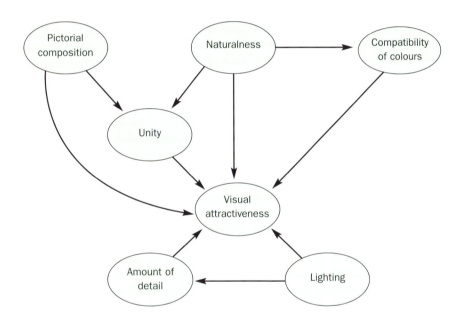

Figure 1: *Relationships between various aspects of the environment and its visual attractiveness. (Taken, with some modification, from a paper by Helliwell (1978) in* **Landscape Research** *3,3)*

Figure 2: *Pedestrian shopping street, with sufficient detail, light and interest to be attractive without any trees. The hanging baskets and window boxes containing flowers provide some element of naturalness and detail.*

Continuing Professional Development

Figure 3: Building without sufficient detail or interest. A few more trees would help to create a more civilised environment.

Figure 4: Other things, besides trees, can contribute to the landscape.

Guidance Notes
©Arboricultural Association

2. Trees and woodlands in the landscape

Fig. 1 illustrates some of the main relationships between various aspects of the environment and its visual attractiveness. It is possible (e.g., Fig. 2) to have attractive landscapes and townscapes in which there are a few or no trees; but trees can often enhance the scene in a variety of ways, including:

- increasing the **unity** of a landscape, by obscuring or dominating miscellaneous items such as parked cars, sheds, and signs

- adding an element of **naturalness** to otherwise "artificial" scenery

- providing interesting **detail** in areas composed of very simple structures or patterns

- enhancing the **composition** of the landscape, whether designed or not

- catching the **light** and creating interest in an otherwise dull scene

- providing a green background to a variety of colours, and thereby **reducing discordancy**

- **providing a distinctive element**, by the choice of particular species or grouping of trees.

It also follows that trees planted in the wrong place can **detract** from the landscape, as can trees which are of an inappropriate colour, shape or density. Such trees may have no positive visual amenity value. [They may have a negative value, but we have not attempted to quantify that.]

Fig. 5:   Urban scene lacking any natural features, and not very "comfortable" or "human".

Continuing Professional Development

3. Monetary value

As has already been outlined, the basis of the Helliwell system is the comparison of different trees or woodlands on a common scale of relative value, in terms of their visual contribution to the landscape. The attachment of a monetary value to this scale may not always be essential, but only by being given such a value can trees and woodlands be assessed and managed on an equal footing with other assets. If the amenity value of trees is excluded from the balance sheet they will be excluded from proper budgeting consideration and their planting, maintenance, and management may receive too much or too little emphasis, according to the interests or whim of the person or body which controls funding.

There is no simple straightforward method of attaching a monetary value to visual amenity, but there are several methods in existence.

The method described in this publication enables the assessor to ascribe a visual amenity value to a tree or woodland on a points scale. This figure can then be multiplied by a conversion factor, to arrive at an appropriate monetary value for planning purposes. The important features of such a method are:

(i)   *the system used to arrive at a points value must be soundly based,*
(ii)   *the intervals on the scale must have similar values,*
(iii)   *the monetary conversion factor must be realistic, and must be accepted as such by a wide spectrum of users, in addition to being clearly independent of any particular case.*

After consultation within and outside the Arboricultural Association, a conversion factor of £10 was recommended for individual trees and £25 for woodlands, as at 1 January 1990. Since then, these figures have been amended to take account of changes in the value of the £. A new appraisal was required following the changes that have been made in this revision, and this has been undertaken by the working group responsible for the revision. As at 1st June 2008, 1 unit = £25 for individual trees, and for woodlands 1 unit = £100. Annually updated figures will be posted on the Arboricultural Association website.

These figures will need to be re-assessed *de novo* from time to time, to ensure that they remain realistic. This assessment may be made with reference to property prices, the value of the tourist trade, effects on mental health and well-being, or instances where marginal decisions have been taken, by public bodies, private companies, or individuals, on the amount of money that could appropriately be spent on retaining, planting, or managing trees and woodlands. This could either be related to extensive studies and/or (as has happened so far) by taking a consensus among a range of informed people. This latter course is simpler and may, in some respects, be more robust. The emphasis, to date, has been on obtaining a conversion factor which is clearly *independent* of any particular case or local pressure group, and which gives results which are *realistic and appropriate*. However, it is expected that additional studies will be undertaken within the next few years, to link the values obtained by this system to other items which have more widely accepted monetary values.

Continuing
Professional
Development

4. Evaluation method for trees

Six factors are identified for each tree. Special factors such as historical association or exceptional rarity, which are not purely visual, were also taken into account originally. However, it is now considered that these did not fit readily into the basic system, and could also be misapplied, so they are no longer included. Other "special" factors such as the screening of unpleasant views or the importance of a tree within a larger composition should be assessed within the six basic factors.

For each of these factors the tree is given a score, and the scores for all six factors are then multiplied together. It should perhaps be stated here that the multiplication of the scores is a necessary procedure where factors are inter-related but act independently, as is the case here. A tree which is large, long-lived, prominent, and suitable for its setting will have a high score. If, on the other hand, a tree is very small, so defective as to have little further useful life expectancy, or is totally unsuited to its setting, it will score little or nothing under the relevant heading and when the scores are multiplied together the tree will have little or no overall score, even if it has scored highly on some other factors. An additive system would not achieve this result.

The product of the scores may then be multiplied by the agreed monetary conversion factor to arrive at an assessment of the visual amenity value of the tree in monetary terms.

Explanatory notes of these factors are given on pages 12 to 18.

Factor	Points									
	0	0.5	1	2	3	4	5	6	7	8
i. Size	Less than 2m²	2 - 5m²	5 - 10m²	10 - 20m²	20 - 30m²	30 - 50m²	50 - 100m²	100 - 150m²	150 - 200m²	over 200m²
ii. Duration	Less than 2 years		2 - 5 years	5 - 40 years	40 - 100 years	100+ years				
iii. Importance	None	Very little	Little	Some	Considerable	Great				
iv. Tree cover		Woodland	Many	Some	Few	None				
v. Suitability to setting	Not	Poor	Just	Fairly	Very	Particularly				
vi. Form		Poor	Average	Good						

Figure 6: Visual amenity valuation table, showing factors and scores available for individual trees. (Refer to following text for details.)

The question of the visual amenity value (as distinct from the biological value) of dead trees has been raised recently. The presence of dead trees may contribute to the character of the landscape. Traditionally, dead trees have been regarded by most people as a sign of neglect. However, artists and designers of the 18th and 19th Centuries, in particular, recognised their value, and there is increasing interest in such trees, particularly if they are very old or large. A veteran oak or sweet chestnut, in addition to harbouring a range of insects, fungi, birds, and possibly bats, may continue to possess some visual value for many years after its death. This value is unlikely to be as great as that of a veteran tree which is still alive, and will be reflected in its reduced size and expected duration, although it may score highly on some other factors.

The assessment of some of the six factors can be carried out with more or less complete objectivity. For example, the size of the tree can be measured fairly simply if the advice in this publication is followed, and percentage tree cover could be measured from aerial photographs (which are readily available for most areas, *via* the internet). However, with some of the factors there will inevitably be a degree of subjectivity, and it may be that two people will assess the same tree differently; or it may even be that the same person will arrive at a slightly different assessment on different days. The amount of disagreement in most cases should not be very great and it is likely that the scope of disagreement would be a matter of one person saying that a particular tree should score 2 and another saying that it should score 3 under a particular factor.

In some cases, the use of half-point scores (1.5, 2.5, 3.5) may be useful, where trees do not fall clearly into any of the main divisions.

Some typical examples of the use of this method of valuation are given on pages 23 to 34.

Explanatory notes

i. Size of tree

The size of the tree is assessed as the area of the tree when viewed from one side. (If this varies from one viewpoint to another, an average figure should be taken.)

The height should be accurately assessed, or measured with a hypsometer or clinometer, and multiplied by the average crown diameter. The stem of the tree should be included in this measurement.

In the examples illustrated below, each small square represents 4m²:

Tree a: is 22.5m tall and the calculated area is 135m². The average crown diameter is 6m and is indicated by shading on the drawing.

Tree b: is 20m tall, has an average crown diameter of 12m, and a calculated area of 240m².

Tree c: is 30m tall, has an average crown diameter of 4m, and a calculated area of 120m².

The crown diameter should be averaged over the full height of the tree; or alternatively, the height and diameter of the crown itself should be estimated, and the area represented by the stem included if it is of significant size, as in **d**.

Tree e: is a veteran tree with a large stem, relatively small crown, and some dead branches, and has a crown area totalling 72m². In a case such as this, in particular, it may be preferable to lay a grid of dots over the sketched outline of the tree and to count the number of dots which coincide with the crown, stem, or branches, as this will have a better chance of recording branches which project from the main crown.

Continuing
Professional
Development

Scoring should be as follows:

		Score
Less than 2m²		Score 0
2 - 5m²	(very small)	Score 0.5
5 - 10m²	small	Score 1
10 - 20m²		Score 2
20 - 30m²		Score 3
30 - 50m²	medium	Score 4
50 - 100m²		Score 5
100 - 150m²	large	Score 6
150 - 200m²		Score 7
200m² ⁺	very large	Score 8

Table 1.   Scores for size.

Trees with a calculated area of less than 2m² should not be valued by this system.

Where two or more trees are growing close together and form a single visual unit they may be valued individually or as one "tree". The maximum size of group which should be assessed as a single unit should be 300m².

Trees a, b, c, d and e.

In examples, 1 square = 4m²

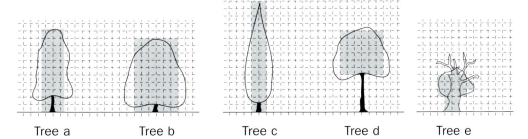

Tree a   Tree b   Tree c   Tree d   Tree e

Diagram 1: Diagram showing tree size calculations for trees with varying canopy shapes.

ii. Expected duration of visual amenity

An estimate should be made of the probable length of time that the tree is likely to contribute to the visual amenity of its location. This will take into account the normal biological life span of trees of that species, its current approximate age, and any factors which may be expected to extend or reduce its life expectancy. For trees which are likely to remain for some time as standing deadwood, a separate valuation based on a reduced *Size* for an additional length of time could be added. This would require some form of discounting of this additional future value; and the uncertainties over a) the life expectancy before the tree dies, b) the length of time that the dead tree is likely to remain standing, c) the rate at which it is likely to reduce in size, and d) the appropriate discount rate will make this a rather uncertain and complicated procedure; which is unlikely to be used for most trees.

It should be pointed out that, while some account is taken of the further duration of the tree, any future changes in its size and condition are not. In addition, the scoring is skewed to some extent in favour of the near future rather than the more distant future, in order to account for the greater uncertainty of the distant future and also the greater imperative to retain trees for the next few years rather than for 50 or 100 years from now. This is partly a reflection of the normal preference to have benefits now rather than later and also the fact that it would be easier to grow trees which would be able to act as significant replacements in 50 years time than in 5 years. This assessment is, therefore, based firmly in the present, but with some acknowledgement to the future.

"Biological" life expectancy. Although some trees can survive for several centuries or (in extreme cases) several thousands of years, most trees have a life expectancy of no more than 300 years under typical conditions in gardens, parks, or streets; and some tree species are unlikely to survive for more than 50 or 60 years. In areas such as large parks with relatively few people and where conditions are suitable for the species in question, some trees may be able to grow for perhaps twice as long, or more.

As a very general guide, some of our common tree species can be grouped into the following categories of biological life expectancy, under typical conditions in parks and gardens in lowland Britain and without any major pruning or other treatment:

350 years or more	*Yew*
250 - 350 years	*Common Oak, Sweet Chestnut, London Plane, Sycamore, Limes, Scots pine*
150 - 250 years	*Cedar of Lebanon, Hornbeam, Beech, Tulip Tree, Norway maple, Corsican pine, Common ash*
100 - 200 years	*Norway Spruce, Walnut, Red Oak, Horse Chestnut, Field Maple, Monkey Puzzle, Mulberry, Pear*
60 - 100 years	*Rowan, Whitebeam, Apple, Wild Cherry, Catalpa, Robinia, most Poplars, Willows, Cherries, Alders and Birches*

Table 2:   Typical biological life expectancy of common tree species in urban conditions.

Continuing
Professional
Development

There will of course be exceptions to this list, and **it should be emphasised that for the purpose of this assessment each tree should be judged individually**. The above figures are intended to provide background information only, and should not be taken as fixed. Trees do not die or fall over due to age, but due to their physiological or structural condition. Trees which have been regularly pollarded from a young age may survive for longer than this; and trees which have been ill-advisedly or poorly pruned or damaged are likely to survive for less.

Safe Useful Life Expectancy. Some account must also be taken of the position in which the tree is growing. For example a tree which overhangs a busy road must be generally sound, yet it may suffer from a variety of ill effects from vehicles or road-works. It would therefore have a lower safe useful life expectancy than a similar tree in a large area of parkland, where the loss of a few branches in a gale would do little harm and where the tree is less likely to suffer damage from road salt, stray vehicles or leaking gas mains. Structural defects may also reduce the safe useful life expectancy of a tree, although it may be possible to correct or alleviate some defects by appropriate surgery. In addition, the level of risk may change if the use of the area near to the tree changes.

Trees which are likely to be removed within 2 years would not normally be valued by this system.

Expected duration	Score
Less than 2 years	0
2 - 5 years	1
5 - 40 years	2
40 - 100 years	3
100+ years	4

Table 3: Scores for expected duration.

iii. Importance of position in landscape

This factor is an expression of the visual prominence of the tree. A single prominent tree in a city centre will rate highly under this factor, while a tree in a remote and secluded area would not.

It could be argued that trees have no *intrinsic* visual amenity value; and trees which can not be seen from any generally accessible public or private vantage point will have little or no actual visual amenity value (although they may, of course, have other values and may have potential visual amenity value if changes occur in the area around them).

If circumstances change which would make a tree more visible, as in the example in Figure 8, a separate calculation of value could be made for the changed circumstances.

The following guidelines should be followed when assessing the visual amenity value of a tree to society as a whole. That is from the point of view of public benefit. The importance of the tree's location is a combination of its prominence in the landscape and the numbers of viewing population.

Continuing Professional Development

"No importance"	=	trees which cannot be seen from any normal vantage point	Score 0
"Very little importance"	=	trees which can only be seen with difficulty or by a very small number of the general public	Score 0.5
"Little importance"	=	most trees in woodlands, back gardens, or in groups of trees etc.	Score 1
"Some importance"	=	individual roadside trees, trees close to busy transport routes, trees in public parks, close to public footpaths, in grounds of hospitals, colleges etc.	Score 2
"Considerable importance"	=	prominent individual trees in well-frequented places such as town centres, village greens, shopping centres etc.	Score 3
"Great importance"	=	trees which are of crucial importance as the principal feature of a public place	Score 4

Table 4:   Scores for importance of position, to the general populace.

Where the amenity to an individual person, family, or group of people is being assessed, the scores allotted should be:

"Little importance"	=	Trees on remote parts of large country estates	Score 0.25
"Some importance"	=	garden trees in groups of no particular individual importance	Score 0.5
"Considerable importance"	=	prominent garden trees	Score 0.75
"Great importance"	=	trees providing a main feature or focal point	Score 1

Table 5:   Scores for the importance of position, to individuals.

It should be noted that, although the maximum score for the amenity value to individuals or small groups is less than the maximum for amenity value to the wider population, some trees may be valued more highly if they are prominent to one or two individuals but are scarcely seen or invisible to the wider public.

iv. Presence of other trees

This factor may overlap to some extent with the previous one, but concentrates on the general abundance of trees within the locality. In areas with a super-abundance of trees, the loss of one of them will, other things being equal, be less important than the loss of a similar tree in an area with few other trees. In fact, in areas which are very densely wooded, the loss of a few trees may be an advantage in visual terms.

Woodland	=	more than 70% of the visual area covered by trees, and at least 100 trees in total	Score 0.5
Many	=	more than 30% of the visual area covered by trees, and at least 10 trees in total	Score 1
Some	=	more than 10% of the visual area covered by trees, and at least four trees in total	Score 2
Few	=	less than 10% of the visual area covered by trees, but at least one other tree present	Score 3
None	=	no other trees present in the area under consideration	Score 4

Table 6: Scores relating to the presence of other trees.

The "visual area" in which the tree is growing is not always easy to define. Sometimes an area such as a town square is surrounded by tall buildings and forms a clearly defined "visual area", but elsewhere there may be no clear definition. In these instances the assessor must make his own decisions and make a note of this. The same definition will be implicit under factors iii and v. For example, if under factor iii a tree is assessed as a prominent tree in a shopping centre, the assessment of the presence of other trees and the relationship of the tree to its setting should relate to the shopping centre, and not to any other areas which may be slightly visible between the buildings.

Where there are no clear boundaries to the "visual area", it is suggested that an area of 50 hectares (400m radius) should be assessed.

The assessment of the percentage cover of trees within the "visual area" should be based on the proportion of the ground in that area which is covered by trees (i.e. as if assessed from an aerial photograph of the area). It is **not** intended to represent a visual impression as seen from ground level.

v. Relation to the setting

This is probably the most difficult factor to determine. As a very general approach, one might aim to have the largest and densest tree, or group of trees, that the available space will appropriately contain [and the word "appropriately" should be emphasised]. Thus a small tree in a large space may appear insignificant. Equally a large tree in a small space may appear overwhelming or "inappropriate". Where trees are close to buildings, light-foliaged trees such as birch or false acacia may be more suitable than dense foliaged species such as beech or sycamore, as they will block out less light from the buildings, cause less of an obstruction to the view, and will complement the appearance of the buildings rather than hiding them. Where trees stand well clear of buildings, however, their visual impact is likely to be greater if they are of a more dense-foliaged type. However, other solutions may be successful in particular instances.

Sometimes a tree, or group of trees, is particularly suitable to a certain setting, in reflecting or emphasising the surrounding landform or architecture. Weeping willows hanging down into flowing water, cedars of Lebanon on a large lawn in front of a mansion, or a row of oak trees in a country lane are examples of suitable juxtapositions. Such trees look right in those places.

Continuing Professional Development

Trees which are screening unpleasant views may be given a higher score under this factor, and trees which have strong autumnal colours or conspicuous flowers might be given a higher rating on that account, but any adjustment to the score should reflect the fact that autumn colours and flowers will only be present for a limited part of the year.

Trees which are part of a formal composition, such as avenues or clumps represent something of a problem in valuation [just as they often do in management terms]. It could be argued that each tree in such a composition should be given a higher score, because its loss would detrimentally affect the whole composition. However, if each tree were treated in this way, that would imply that a formal composition is of greater overall value than an informal grouping of trees; which may sometimes be the case, but is not necessarily so. It is suggested that no additional score should automatically be given when valuing the *presence* of a tree within a formal composition, but when assessing the effects of *loss* of the tree it might also be appropriate to take into account any consequent loss of value of the other trees within the formal composition.

Suitability to setting	Score
Totally unsuitable (Much too large, much too small, obscuring attractive view, disrupting formal composition, totally wrong colour, etc. Landscape would be improved if tree removed.)	Score 0
Moderately unsuitable	Score 0.5
Just suitable	Score 1
Fairly suitable (Fairly well placed. A definite asset to the landscape.)	Score 2
Very suitable (Well placed *or* screening unsightly views.)	Score 3
Particularly suitable (Well placed *and* screening unsightly views, *or* making a special contribution to local character.)	Score 4

Table 7: Scores for relation to setting.

vi. Form

The form of a tree is, again, a matter which may be difficult to define precisely, although extreme examples of badly mutilated trees, or trees which have suffered damage from disease or storms will be fairly obvious. Form need not necessarily be natural to be good: a good example of a pollarded tree may merit a high score; and deformed trees are not necessarily totally worthless. **Most trees will be rated "average".** Trees which are rated higher than average may have stems which are much thicker than normal for the size of the crown, a particularly gnarled branch form, attractive pendulous branches, or a markedly windswept appearance (when growing in exposed coastal locations, for example). **It should be emphasised that form is being assessed here in entirely aesthetic terms, and is not related to the structural condition of the tree (which may affect its longevity and/or the cost of maintaining it;** both of which may need to be taken into account, either within or as a subscript to the valuation.)

Form	Score
Poor	Score 0.5 - 0.9
Average/indifferent	Score 1
Good	Score 1.1 - 2

Table 8: Scores for form.

Continuing Professional Development

5. Evaluation method for woodlands

Six factors are identified for each woodland area. For each of these factors the woodland is given a score, and the scores for all the factors are then multiplied together to give an assessment of the amenity value of the woodland.

If a woodland is less than 0.1 hectare it should be evaluated as a group of individual trees (or as one very large "tree") and not as woodland, using the Evaluation Method for Trees.

Woodlands which (due to their position, shape or other reason) are unacceptable in the landscape should score zero under factor vi. This is not to say that a poorly-shaped woodland could not be an amenity if it were modified in some way, and it should be stressed that this method needs to be applied sensibly, with due regard to such matters. It may, for example, be possible to assess the amenity value of a woodland in 20 years' time, on the assumption that an additional area is planted now, and to use that assessment to arrive at a decision on the awarding of grant aid for such planting. With several of the factors there will be an element of subjectivity, and there may be some disagreement between different people's assessments, as has already been discussed in Section 4.

Where trees are growing either as isolated individual trees or as dense woodland it will normally be quite clear as to which evaluation procedure should be followed. However, in places where there are numerous trees, clumps, or groups in close proximity, or where there is woodland with numerous gaps, it may be less clear. It may be possible to value such trees or woodlands either as numerous single trees or as a single area of woodland. Where it is not clear which category is most appropriate, it will usually be more practical to assess the area as woodland, particularly if the trees and the areas between them form a single unit of land management, such as wood pasture.

Factor	Points				
	0.5	1	2	3	4
i. Size of woodland		Very small	Small	Medium	Large
ii. Position in landscape	Very secluded	Secluded	Visible, but not prominent	Prominent	Very prominent
iii. Viewing population	Very few	Few	Some	Many	Very many
iv. Presence of other trees and woodland in the vicinity	Surrounding area more than 75% wooded	Surrounding area more than 25% wooded	Surrounding area 5 - 25% wooded	Surrounding area 1 - 5% wooded	Surrounding area less than 1% wooded
v. Composition and structure of the woodland	Plantation with geometric stripes, or visually degraded woodland	Even-aged young woodland	Mature or uneven-aged woodland or wood pasture	Mature or uneven-aged woodland or wood pasture with large or veteran trees	
vi. Compatibility	Only just acceptable	Acceptable	Moderately good	Good	Excellent

Table 9: Scores available for woodlands.

Again, this method of assessment only takes account of the *visual amenity value* of a woodland, and does not assess other benefits, or the costs involved in managing the woodland. Such costs are usually ascertainable, and will need to be considered before management decisions are taken.

Some typical examples of the use of this method of valuation are given on pages 35 to 39.

Explanatory notes

i. The size of the woodland being evaluated

Woods in excess of 40 hectares should be evaluated in more than one unit; and where different parts of a wood are of different quality or visibility these should also be evaluated separately. Ownership boundaries will not normally be relevant, if they are not coincident with changes in appearance or visibility.

The relevant points score can be taken from the following scale and the values for other sizes interpolated.

size of woodland (ha)										
0.1	0.25	0.5	1	2	4	8	15	20	30	40
points										
0.75	1	1.5	2	2.5	3	3.5	3.8	4	4.3	4.5

Table 10: Scores for size.

Where, due to the topography, only the outer edge of the woodland is visible, it is suggested that an area equivalent to a band 50m wide should be assessed for its contribution to the landscape. If there is public access within other parts of the woodland, this may need to be assessed separately.

ii. Position in landscape

A *"very secluded"* woodland is one which is difficult to see from anywhere outside the woodland.

A *"secluded"* woodland is one which is visible from a restricted area of not more than about one square kilometre.

A *"non-prominent"* woodland will be visible, but not particularly prominent, over an area of one to four square kilometres.

A *"prominent"* woodland will be **readily** visible over an area of at least two square kilometres;

A *"very prominent"* woodland will be prominently visible over an area of more than four square kilometres.

Very secluded	=	Difficult to see	0.5
Secluded	=	Visible from less than 1km²	1
Non-prominent	=	Visible from more than 1km²	2
Prominent	=	Readily visible from less than 2km²	3
Very prominent	=	Prominent for less than 4km²	4

Table 11: Scores for position.

It should be noted that the above descriptors are internal to this document only. For a wider discussion and guidance on visual impact assessment see Landscape Institute (2002).

Continuing
Professional
Development

Guidance Notes
©Arboricultural Association

iii. Calculations of **"viewing population"** are unlikely to be exact. A sensible procedure may be to take a relatively small percentage (say 2%) of the resident population who live in houses within sight of the woodland; plus a percentage of people in vehicles on nearby roads (say 25%) at any one time during average daytime traffic conditions; plus the average numbers of people on foot who are likely to be within viewing distance at any one time. Thus, a woodland on the edge of a town, within sight of about 300 houses (containing about 1,000 people); visible from about 2km of motorway; and from several footpaths and play areas may have an average "viewing population" of 20 + 50 + 30 = 100.

Using this framework, the following figures and scores are suggested for the various categories:

Very few	=	less than 1 average viewing population	0.5
Few	=	1 - 2 viewing population	1
Some	=	2 - 20 viewing population	2
Many	=	20 - 100 viewing population	3
Very many	=	100+ viewing population	4

Table 12: Scores for viewing population.

This categorisation is preliminary and would be capable of further refinement, possibly (but not necessarily) differentiating between residents who are retired from those who are of working age; people travelling for business purposes from holidaymakers; and people who are walking for pleasure from those walking to work or school. The figures suggested above are only a first attempt to give an appropriate weighting, in the absence of better information.

iv. Presence of other trees and woodland

This factor is included to allow for the decreasing visual importance of any one individual area of woodland as the amount of other trees and woodland in the vicinity increases. (Maximum overall amenity values for a parish or district are often obtained with 50 to 60% woodland cover).

The area which should be assessed for tree and woodland cover should normally be four square kilometres.

(The woodland being assessed should not be included in the percentage figure.)

Area more than 75% wooded	0.5
More than 25% wooded	1
5 - 25% wooded	2
1 - 5% wooded	3
Less than 1% wooded	4

Table 13: Scores for presence of other woodland.

v. Composition and structure of the woodland

This is intended to be a purely visual assessment, giving young or monotonous plantations a low score and diverse or mature woodland a higher score. A pure even-aged spruce plantation is likely to be classed as monotonous and uninteresting at almost any stage, unless it is allowed to grow on past the normal age for commercial felling, and will score no more than 1; but a plantation containing mature pines, with a mixture of birch and with some open gaps and a ground vegetation of heather and bilberry may be very attractive and score 3.

In the right place and with appropriate management, a plantation of almost any species could score highly. It should also be noted that woodlands which are managed on an uneven-aged basis and are not likely to be clear felled at some stage are likely to retain their visual amenity value in the longer term. However, it is assumed that woodland will continue to be woodland, and the valuation derived by this system is focussed primarily on the current composition and structure.

Plantation with geometric stripes, or visually degraded woodland	0.5
Even-aged young woodland	1
Mature or uneven-aged* woodland or wood pasture**	2
Mature woodland, uneven-aged woodland, or wood pasture with very large or veteran trees	3

Table 14: Scores for composition.

*   ideally, uneven-aged woodland will include approximately equal proportions of small, medium, and large sized trees
**   wood pasture is woodland of an open character which is used by grazing animals on a regular basis.

vi. Compatibility in the landscape

As noted above, woodlands which fit very badly into the landscape should receive a zero score. Examples of such woodlands might be rectangular shelterbelts or plantations running contrary to the lie of the land on hillsides; where the landscape would be visually improved if they were removed.

Woodlands which are "only just acceptable" might include coniferous plantations in a predominantly broadleaved landscape, or woodlands of any type which are of a rigid or awkward shape. In most cases it would be easy to envisage ways in which a modest change in species, management, or shape of the woodland would improve its appearance.

Woodlands which bear little relation to the wider landscape but are not clearly inappropriate will probably be "acceptable", and woodlands which fit in reasonably or very well within the landform and other land uses may be rated "moderately good" or "good".

"Excellent" ratings might be given to woods which fit so well that it is difficult to imagine the landscape without them.

Not acceptable	0
Only just acceptable	0.5
Acceptable	1
Moderately good	2
Good	3
Excellent	4

Table 15: Scores for compatibility.

Continuing
Professional
Development

Guidance Notes
©Arboricultural Association

6. EXAMPLES

Fig. 7: Small trees planted in containers in motorway services area. (They are held upright by guy wires which run from the edge of the soil area to a point above the lowest branches, and are kept alive by regular (or irregular) watering, but do not appear to be thriving.)

The value of one of these trees could be assessed as follows:

1.	*Size*	*4m²*	*scores 0.5 points*
2.	*Expected duration*	*around 15 years*	*scores 2 points*
3.	*Importance in landscape*	*in public space*	*scores 3 points*
4.	*Presence of other trees*	*less than 10%*	*scores 3 points*
5.	*Relation to setting*	*fairly suitable*	*scores 2 points*
6.	*Form*	*average*	*scores 1 point*

Total score = 0.5 x 2 x 3 x 3 x 2 x 1 = 18, or £450

Fig. 8: Row of lime trees at rear of existing houses and on edge of impending development.

Before the new houses are built, the value of a typical tree in this row would be:

1.	*Size*	*50m²*	*scores 4.5 points*
2.	*Expected duration*	*around 60 years*	*scores 3 points*
3.	*Importance in landscape*	*not very great*	*scores 1 point*
4.	*Presence of other trees*	*more than 10%*	*scores 2 points*
5.	*Relation to setting*	*very suitable*	*scores 3 points*
6.	*Form*	*average*	*scores 1 point*

Total score = 4.5 x 3 x 1 x 2 x 3 x 1 = 81, or £2,025, or around £30,375 for the whole row

After construction of the new development, with these trees in a relatively prominent location adjacent to a road, each tree is likely to be valued at:

1.	*Size*	*50m²*	*scores 4.5 points*
2.	*Expected duration*	*around 60 years*	*scores 3 points*
3.	*Importance in landscape*	*some*	*scores 2 points*
4.	*Presence of other trees*	*more than 10%*	*scores 2 points*
5.	*Relation to setting*	*very suitable*	*scores 3 points*
6.	*Form*	*average*	*scores 1 point*

Total score = 4.5 x 3 x 2 x 2 x 3 x 1 = 162, or £4,050, or around £60,750 for the whole row

Fig. 9: Group of young birch trees close to new houses, which will form part of a larger belt, augmented by planting, within an open space.

These may be valued as one large "tree":

1.	*Size*	*120m²*	*scores 6 points*
2.	*Expected duration*	*around 40 years*	*scores 2.5 points*
3.	*Importance in landscape*	*roadside*	*scores 2 points*
4.	*Presence of other trees*	*more than 10%*	*scores 2 points*
5.	*Relation to setting*	*very suitable*	*scores 3 points*
6.	Form	*average*	scores 1 point

Total score = 6 x 2.5 x 2 x 2 x 3 x 1 = 180, or £4,500

Fig. 10: Pruned plane trees.

These trees would look less ugly in the summer, with some new growth, but even then they would be of less amenity value than trees of more natural form. In winter, they would probably be judged to be only barely suitable for their setting, giving a relatively low value:

1.	Size	$9m^2$	scores 1 point
2.	Expected duration	around 40 years	scores 2.5 points
3.	Importance in landscape	roadside tree	scores 2 points
4.	Presence of other trees	more than 10%	scores 2 points
5.	Relation to setting	moderately unsuitable	scores 1 point
6.	Form	poor	scores 0.5 points

Total score = 1 x 2.5 x 2 x 2 x 1 x 0.5 = 10, or £250

Fig. 11: Avenue of lime trees, lining footpath on rural estate.

The trees are closely spaced and form a linear feature as seen from the footpath, but are surrounded by woodland on both sides, so do not register very strongly in the broader landscape. The loss of any one tree would not greatly affect the overall effect.

The value of one of a typical tree would be:

1.	*Size*	*60m²*	*scores 5 points*
2.	*Expected duration*	*around 40 years*	*scores 2.5 points*
3.	*Importance in landscape*	*little/some*	*scores 1.5 points*
4.	*Presence of other trees*	*more than 70%*	*scores 0.5 point*
5.	*Relation to setting*	*fairly/very suitable*	*scores 2.5 points*
6.	*Form*	*average*	*scores 1 point*

Total score = 5 x 2.5 x 1.5 x 0.5 x 2.5 x 1 = 23, or £575 per tree

Continuing Professional Development

Fig. 12: Purple plum.

Rather close to windows of house which is set back from the others, but contributes to the street scene. The purple colour may not blend in well with the red brick houses. A green tree might have been better in this respect.

1.	*Size*	*12m²*	*scores 2 points*
2.	*Expected duration*	*around 25 years*	*scores 2 points*
3.	*Importance in landscape*	*individual roadside tree*	*scores 2 point*
4.	*Presence of other trees*	*more than 10%*	*scores 2 points*
5.	*Relation to setting*	*fairly suitable*	*scores 2 points*
6.	*Form*	*average*	*scores 1 point*

Total score = 2 x 2 x 2 x 2 x 2 x 1 = 32, or £800

Fig. 13: A row of very small Lawson cypress has been planted close to the chain-link fence.

Fig. 14: Seven years later, the cypress trees have grown to form a screen, hiding the clutter of vehicles, sheds, etc.

This row of cypress trees can be evaluated as one "tree", as follows:

1.	*Size*	*40m²*	*scores 4 points*
2.	*Expected duration*	*around 30 years*	*scores 2 points*
3.	*Importance in landscape*	*some*	*scores 2 points*
4.	*Presence of other trees*	*less than 10%*	*scores 3 points*
5.	*Relation to setting*	*particularly suitable*	*scores 4 points*
6.	*Form*	*average*	*scores 1 point*

Total score = 4 x 2 x 2 x 3 x 4 x 1 = 192, or £4,800

Continuing Professional Development

Fig. 15: Clipped yew trees in a Gloucestershire churchyard. These are a well-known feature of this particular churchyard.

There are approximately 100 clipped yews in this churchyard and some other trees, in the centre of a picturesque Cotswold village.

Each of the yews might be valued at:

1.	*Size*	*7m²*	*scores 1 point*
2.	*Expected duration*	*around 60 years*	*scores 3 points*
3.	*Importance in landscape*	*some*	*scores 2 points*
4.	*Presence of other trees*	*more than 30%*	*scores 1 point*
5.	*Relation to setting*	*very suitable*	*scores 3 points*
6.	*Form*	*fair/good*	*scores 1.5 points*

Total score 1 x 3 x 2 x 1 x 3 x 1.5 = 27, or £675

Although yew trees can survive for hundreds of years, these clipped yews appear to be less long lived, and one or two trees die every few years.

This would place a value of £67,500 on 100 such trees.

Continuing Professional Development

Fig. 16: Row of weeping willows between a factory and a canalised river.

These trees make a considerable difference to this scene.

1.	*Size*	*$60m^2$*	*scores 5 points*
2.	*Expected duration*	*around 30 years*	*scores 2 points*
3.	*Importance in landscape*	*some*	*scores 2 points*
4.	*Presence of other trees*	*less than 10%*	*scores 3 points*
5.	*Relation to setting*	*very suitable*	*scores 3 points*
6.	*Form*	*average*	*scores 1 point*

Each tree scores 5 x 2 x 2 x 3 x 3 x 1 = 180, or £4,500

Fig. 17: Old oak in grassland.

This tree appears to have suffered some die-back as a result of damage to the root system by ploughing, and is developing some of the characteristics of an ancient tree. It is possible that it may have some value for wildlife, although it is not part of a larger population of ancient trees. However, as explained in the Introduction (page 5), we are only considering visual amenity values in this system.

1.	*Size*	*40m²*	*scores 4 points*
2.	*Expected duration*	*100+ years*	*scores 4 points*
3.	*Importance in the landscape*	*little*	*scores 1 point*
4.	*Presence of other trees*	*more than 30%*	*scores 1 point*
5.	*Relation to setting*	*fairly suitable*	*scores 2 points*
6.	*Form*	*average*	*scores 1 point*

Total score 4 x 4 x 1 x 1 x 2 x 1 = 32, or £800

Fig. 18: Lombardy poplars in residential neighbourhood.

This row of 11 trees is prominent in the locality and forms something of a feature, although the stiff formality of this type of tree does not relate very clearly to any other features of this landscape.

1.	*Size*	*40m²*	*score 4 points*
2.	*Expected duration*	*around 20 years*	*score 2 points*
3.	*Importance in landscape*	*some*	*score 2 points*
4.	*Presence of other trees*	*more than 10%*	*score 2 points*
5.	*Relation to setting*	*slightly unsuitable*	*score 0.75 point*
6.	*Form*	*average*	*score 1 point*

Total score = 4 x 2 x 2 x 2 x 0.75 x 1 = 24, or £600 each (or £6,600 for the complete row)

Fig. 19: Tree which has been pruned to give clearance from power lines.

1.	Size	40m²	score 4 points
2.	Expected duration	around 25 years	score 2 points
3.	Importance	some	score 2 points
4.	Presence of other trees	more than 10%	score 2 points
5.	Relation to setting	fairly suitable	score 2 points
6.	Form	poor	score 0.5 points

Total score = 4 x 2 x 2 x 2 x 2 x 0.5 = 32, or £800

Fig. 20: Mixed woodland, clothing a low ridge, visible from road.

Using the evaluation method for woodlands, an area of around eight hectares of such woodland might be valued:

1.	*Size*	*8ha*	*scores 3.5 points*
2.	*Position in landscape*	*prominent*	*scores 3 points*
3.	*Viewing population*	*average*	*scores 2 points*
4.	*Presence of other trees and woodland*	*5-25%*	*scores 2 points*
5.	*Composition and structure*	*mature, even-aged*	*scores 1.5 points*
6.	*Compatibility in the landscape*	*good*	*scores 3 points*

Total score = 3.5 x 3 x 2 x 2 x 1.5 x 3 = 189 or £18,900 (= £2,362 per hectare)

If this woodland were divided into two smaller areas of 3.5 hectares by the removal of a part in the centre, the evaluation of each of the remaining areas might be:

1.	*Size*	*3.5ha*	*scores 3 points*
2.	*Position in landscape*	*prominent*	*scores 3 points*
3.	*Viewing population*	*average*	*scores 2 points*
4.	*Presence of other trees and woodland*	*5-25%*	*scores 2 points*
5.	*Composition and structure*	*mature, even-aged*	*scores 1.5 points*
6.	*Compatibility in the landscape*	*acceptable/ moderately good*	*scores 1.5 points*

Total score = 3 x 3 x 2 x 2 x 1.5 x 1.5 = 81, or £8,100 (= about £2,314 per hectare)

Continuing Professional Development

Fig. 21: Interior of beech woodland, managed under a group selection system.

This would score 2 for composition and structure.

Fig. 22: Dense even-aged spruce plantation which would score 1 for composition and structure.

Continuing Professional Development

Guidance Notes
©Arboricultural Association

Fig. 24: Clear felled conifers

Fig. 23: Uneven-aged spruce, which would score 2 for composition and structure

An area of 20 hectares of this landscape *(Fig. 24)* might be valued:

1.	*Size*	*20ha*	*scores 4 points*
2.	*Position in landscape*	*secluded*	*scores 1 point*
3.	*Viewing population*	*few*	*scores 1 point*
4.	*Presence of other trees and woodland*	*5-25%*	*scores 2 points*
5.	*Composition and structure*	*dense plantation*	*scores 1 point*
6.	*Compatibility in the landscape*	*just acceptable*	*scores 0.5 points*

Total score = 4 x 1 x 1 x 2 x 1 x 0.5 = 4, or £400 (= £20 per hectare)

That is, little more than zero. Some modification of the boundaries of felled areas and choice of species may result in a larger score.

Continuing Professional Development

Fig. 25: Spruce of mixed ages, scoring 2 for composition and structure   *Fig. 26: Uneven-aged forest of mixed species*

An area of 40 hectares *(Fig. 26)*, with public access, but in a relatively remote area might be valued:

1.	*Size*	*20ha*	*scores 4 points*
2.	*Position in landscape*	*secluded*	*scores 1 point*
3.	*Viewing population*	*few*	*scores 1 point*
4.	*Presence of other trees and woodland*	*more than 25%*	*scores 1 point*
5.	*Composition and structure*	*uneven-aged with fairly large trees*	*scores 2.5 points*
6.	*Compatibility in the landscape*	*good*	*scores 3 points*

Total score = 4 x 1 x 1 x 1 x 2.5 x 3 = 30, or £3,000 (= £150 per hectare)

Continuing Professional Development

Fig. 27: Even-aged oak woodland in a popular scenic area.

10 hectares might be valued:

1.	Size	10ha	scores 3.6 points
2.	Position in landscape	average	scores 2 points
3.	Viewing population	average	scores 2 points
4.	Presence of other trees and woodland	5-25%	scores 2 points
5.	Composition and structure	mature	scores 2 points
6.	Compatibility in the landscape	good	scores 3 points

Total score = 3.6 x 2 x 2 x 2 x 2 x 3 = 173, or £17,300 (= £1,730 per hectare)

It may be possible to value this as a number of smaller units, taking into account the varying accessibility, prominence and character of each area. The total value should be approximately the same in each case (in spite of the apparently small decrease in value allotted to smaller areas under item 1).

It is hoped that these examples may serve to illustrate the use of these valuation methods. There will rarely be a completely definitive value for any particular tree or woodland, but a realistic value should be obtained if the methods are applied sensibly and fairly.

Continuing Professional Development

7. References and other systems of amenity valuation of trees

Asociacion Espanola de Parques y Jardines Publicos. 1990. Metado de valoracion del arbolado ornamental: Norma Granada. **A.E.P.J.P. Madrid.**

Barrell, J.D. 1993. Pre-development tree surveys: Safe Useful Life Expectancy (SULE) is the natural progression. **Arboricultural Journal 17, 33-46.**

Barry, G.C. and Murray, T.G. 1982. A monetary evaluation of trees in Ireland. **Irish Landscape Journal 1(5), 5-16.**

Bary-Lenger, A. and Nebout, J-P. 2002. Évaluation financière des arbres d'agrément et de production. **Editions TEC & DOC, Paris**

Council of Tree and Landscape Appraisers 2000. Guide for plant appraisal. (9th edition). **International Society of Arboriculture,** P.O. Box 3129, Champaign, Illinois, 61826-3129, United States of America.

Dolwin, J.A. and Goss, C.L. 1993. Evaluation of amenity trees within the Borough of Tunbridge Wells. **Arboricultural Journal 17, 301-308.**

Helliwell, D.R. 1967. The amenity value of trees and woodlands. **Aboricultural Journal 1, 128-131.**

Helliwell, D.R. 1978. The assessment of landscape preferences. **Landscape Research 3 (3), 15-17.**

Helliwell, D.R. 1983. Garden trees. John Wiley & Sons Ltd., Chichester.

Landscape Institute with the Institute of Environmental Management and Assessment. 2002 (2nd ed.). Guidelines for landscape and visual impact assessment. Spon Press, London.

Lohr, V.I. and Pearson-Mims. 2006. Responses to scenes with spreading, rounded, and conical tree forms. **Environment and behaviour 38, 667-688.**

Lopez Arce, M.A. and Alamo, D. del. 1975. El calculo de indemnizaciones derivadas de la perdida de arboles ornamentales. (Calculating compensation for the loss of ornamental trees). **Boletin de la Estacion Central de Ecologia 4 (7), 3-19.**

Ministerie van Openbare Werken, 1979. Uniforme methode voor waardebepaling van stratt-, laan-, en parkbomen behorend tot het openbaar domein. (Standard method for valuation of street, avenue, and park trees in the public domain.) **Ministerie van Openbare Werken,** Dienst van het Groenplan, Wetstraat 151, 1040 Brussels.

Price, C 1993. Time, discounting, and value Blackwell, Oxford.

Price, C. 2003. Quantifying the aesthetic benefits of urban forestry. **Urban Forestry and Urban Greening 1, 123-133.**

Price, C. 2007. Putting a value on trees: an economist's perspective. **Arboricultural Journal 30, 7-19.**

Raad, A. 1976. Trees in towns and their evaluation. **Arboricultural Journal 3, 2-16.**

Royal New Zealand Institute of Horticulture, 1988. A tree evaluation system for New Zealand. **R.N.Z.I.H.,** P.O. Box 11379, Wellington, New Zealand.

Willis, K.G. and Garrod, G.D. 1993. The contribution of trees and woodlands to the value of property. **Arboricultural Journal 17, 211-219.**

8. Attribution and acknowledgements

The system adopted here is a modified form of proposals by Rodney Helliwell which were published in the Arboricultural Journal in 1967, and may be referred to as the Helliwell System 2008.

This publication supersedes earlier versions published by the Tree Council in 1974 and by the Arboricultural Association in 1984, 1990, 1994, 2000, and 2003.

The text has been revised and extended by Rodney Helliwell, with considerable assistance from Steve Coombes and members of a working group set up by the Tree Council and the Arboricultural Association in 2007. Suggestions made by participants at workshops organised by the Arboricultural Association between 2001 and 2007 have also been helpful.

Photographs in the text have been supplied by Rodney Helliwell, Mike Volp and Bill Anderson.

Continuing
Professional
Development

Data Sheet for Trees

TREE NUMBER				SCORE	NOTES
1.	**Size**				
	0	less than	2m²		
	0.5	very small	2 - 5m²		
	1	small	5 - 10m²		
	2		10 - 20m²		
	3		20 - 30m²		
	4	medium	30 - 50m²		
	5		50 - 100m²		
	6	large	100 - 150m²		
	7		150 - 200m²		
	8	very large	200m²+		
2.	**Expected duration**				
	0	less than 2 years			
	1	2-5 years			
	2	5-40 years			
	3	40-100 years			
	4	100+ years			
3.	**Position (Importance in the Landscape)**				
Private assessment					
	0.25	*Little importance:* trees on remote parts of large country estates			
	0.5	*Some importance:* garden trees in groups of no particular individual importance			
	0.75	*Considerable importance:* prominent garden trees			
	1.0	*Great importance:* main feature or focal point			
Public assessment					
	0	*No importance:* trees not visible from any public vantage point			
	0.5	*Very little importance:* trees only seen with difficulty or by a very small number of people			
	1	*Little importance:* most trees in woodlands, back gardens or in groups of trees, etc.			
	2	*Some importance:* individual roadside trees. Trees close to busy roads. Trees in public parks. Close to public footpaths in grounds of hospitals, colleges etc.			
	3	*Considerable importance:* prominent individual trees in well-frequented places such as town centres, shopping centres, etc.			
	4	*Great importance:* trees which are of crucial importance as the principal feature of a public place			
4.	**Other trees**				
	0.5	more than 70% of the visual area covered by trees, and at least 100 trees in total			
	1	more than 30% of the visual area covered by trees, and at least 10 trees in total			
	2	more than 10% of the visual area covered by trees, and at least 4 trees in total			
	3	less than 10% of the visual area covered by trees, but at least one other tree present			
	4	no other trees present in the area under consideration			
5.	**Relation to setting**				
	0	totally unsuitable			
	0.5	moderately unsuitable			
	1	just suitable			
	2	fairly suitable			
	3	very suitable			
	4	particularly suitable			
6.	**Form**				
	0.5	Trees which are of poor form			
	1	trees of average form			
	2	trees of above average form			

Continuing Professional Development

Guidance Notes
©Arboricultural Association